# TO MEMBERS OF PARLIAMENT AND OTHERS.

## FORTY-FIVE YEARS OF REGISTRATION STATISTICS, PROVING VACCINATION TO BE BOTH USELESS AND DANGEROUS.

### (1885)

BY

ALFRED RUSSEL WALLACE

British Library Cataloguing-in-Publication Data
A catalogue record for this book is available from the
British Library

# Alfred Russel Wallace

Alfred Russel Wallace was born on 8<sup>th</sup> January 1823 in the village of Llanbadoc, in Monmouthshire, Wales.

At the age of five, Wallace's family moved to Hertford where he later enrolled at Hertford Grammar School. He was educated there until financial difficulties forced his family to withdraw him in 1836. He then boarded with his older brother John before becoming an apprentice to his eldest brother, William, a surveyor. He worked for William for six years until the business declined due to difficult economic conditions.

After a brief period of unemployment, he was hired as a master at the Collegiate School in Leicester to teach drawing, map-making, and surveying. During this time he met the entomologist Henry Bates who inspired Wallace to begin collecting insects. He and bates continued exchanging letters after Wallace left teaching to pursue his surveying career. They corresponded on prominent works of the time such as Charles Darwin's *The Voyage of the Beagle* (1839) and Robert Chamber's *Vestiges of the Natural History of Creation* (1844).

Wallace was inspired by the travelling naturalists of the day and decided to begin his exploration career collecting specimens in the Amazon rainforest. He explored the Rio

Negra for four years, making notes on the peoples and languages he encountered as well as the geography, flora, and fauna. On his return voyage his ship, Helen, caught fire and he and the crew were stranded for ten days before being picked up by the Jordeson, a brig travelling from Cuba to London. All of his specimens aboard Helen had been lost.

After a brief stay in England he embarked on a journey to the Malay Archipelago (now Singapore, Malaysia, and Indonesia). During this eight year period he collected more than 126,000 specimens, several thousand of which represented new species to science. While travelling, Wallace refined his thoughts about evolution and in 1858 he outlined his theory of natural selection in an article he sent to Charles Darwin. This was published in the same year along with Darwin's own theory. Wallace eventually published an account of his travels *The Malay Archipelago* in 1869, and it became one of the most popular books of scientific exploration in the 19$^{th}$ century.

Upon his return to England, in 1862, Wallace became a staunch defender of Darwin's landmark work *On the Origin of Species* (1859). He wrote responses to those critical of the theory of natural selection, including 'Remarks on the Rev. S. Haughton's Paper on the Bee's Cell, And on the Origin of Species' (1863) and 'Creation by Law' (1867). The former of these was particularly pleasing to Darwin. Wallace also published important papers such as 'The Origin of Human

Races and the Antiquity of Man Deduced from the Theory of 'Natural Selection" (1864) and books, including the much cited *Darwinism* (1889).

Wallace made a huge contribution to the natural sciences and he will continue to be remembered as one of the key figures in the development of evolutionary theory.

Wallace died on $7^{th}$ November 1913 at the age of 90. He is buried in a small cemetery at Broadstone, Dorset, England.

# PART I. SMALL-POX MORTALITY AND VACCINATION.

Having been led to enquire for myself as to the effects of Vaccination in preventing or diminishing Small-pox, I have arrived at results as unexpected as they appear to me to be conclusive. The question is one which affects our personal liberty as well as the health and even the lives of thousands; it therefore becomes a duty to endeavour to make the truth known to all, and especially to those who, on the faith of false or misleading statements, have enforced the practice of vaccination by penal laws.

I propose now to establish the following four statements of fact, by means of the only official statistics which are available; and I shall adopt a mode of presenting those statistics as a whole, which will render them intelligible to all. These statements are:--

(1.)--That during the forty-five years of the Registration of deaths and their causes, Small-pox mortality has very slightly diminished, while an exceedingly severe Small-pox epidemic occurred within the last twelve years of the period.

(2.)--That there is no evidence to show that the slight decrease of Small-pox mortality is due to vaccination.

(3.)--That the severity of Small-pox as a disease has not been mitigated by vaccination.

(4.)--That several inoculable diseases have increased to an alarming extent coincidently with enforced vaccination.

The first, second, and fourth propositions will be proved from the Registrar-General's Reports from 1838 to 1882; and I shall make the results clear and indisputable, by presenting the figures for the whole period in the form of diagrammatic curves, so that no manipulation of them, by taking certain years for comparison, or by dividing the period in special ways, will be possible.

The diagrams show, in each case, not the absolute mortality but the deaths per million  living, a method which eliminates the increase of population and gives true comparative results.

## Vaccination Has Not Diminished Small-pox.

Diagram I. exhibits the deaths from Small-pox, in London, for every year from 1838 to 1882, while an upper line exhibits the deaths from the other principal zymotic diseases given in the Registrar-General's Annual Summary for 1882, (except Cholera, which is only an occasional epidemic,) namely,--Scarlet fever and Diphtheria, Measles, Whooping Cough, Typhoid and other fevers, and Diarrhœa. A dotted line between these shows the mortality from fevers of the Typhoid class.[1]

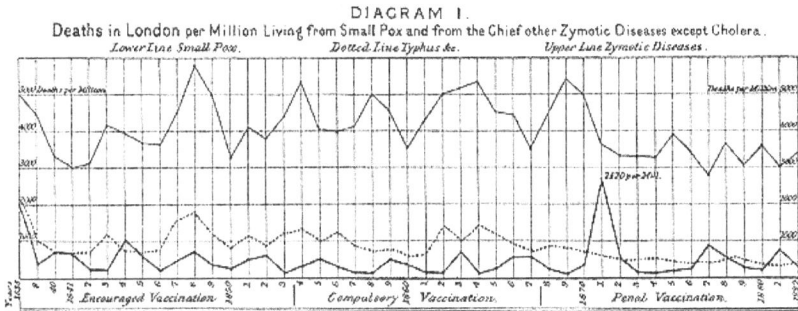

DIAGRAM I.
Deaths in London per Million Living from Small Pox and from the Chief other Zymotic Diseases except Cholera.
*Lower Line Small Pox.*    *Dotted Line Typhus &c.*    *Upper Line Zymotic Diseases.*

The first thing clearly apparent in this diagram, is the very small diminution of Small-pox corresponding with the epochs of penal and compulsory vaccination; while the epidemic of 1871 was the most destructive in the whole period. The average diminution of Small-pox mortality from the first to the second half of the period, is 57 deaths per million per annum. Looking now at the upper curve, we see that the mortality from the chief zymotic diseases has also decreased, more especially during the last 35 years; but the decrease of these diseases is not, proportionally, so great, owing to the fact that deaths from Diarrhœa have considerably increased in the latter half of the period. On the other hand, Typhus and Typhoid fevers have diminished

9

to a much greater extent than Small-pox, as shown by the dotted line on the diagram, the reduced mortality from this cause alone being 382 per million, or more than six times as much as that from Small-pox. Every one will admit that this remarkable decrease of Typhus, &c., is due to more efficient sanitation, greater personal attention to the laws of health, and probably also to more rational methods of treatment. But all these causes of amelioration have certainly had their effect on Small-pox; and as the mortality from that disease has not equally diminished, there is probably some counteracting cause at work. So far, therefore, from there being any proof that vaccination has diminished Small-pox in London, the tendency of the Registrar-General's facts, (and there are no other facts which are trustworthy) is to show that some counteracting cause has prevented general sanitation from acting on this disease as it has acted on Typhus, and that cause may, possibly, be vaccination itself.

We will now turn to Diagram II., which gives a representation of similar statistics for England and Wales,[2] except that unfortunately there is a blank in the record for 1843-46, in which years the Registrar-General informs us, "the causes of death were not distinguished." Here too we perceive a similar decrease in Small-pox mortality, broken by the tremendous epidemic of 1871-2, while the other chief zymotic diseases represented by the higher line, show more irregularity, but a considerable recent decrease.

For all England, as for London, the tables show us that Typhoid fevers have decreased far more than Small-pox, (but for clearness the curve of Typhus is omitted,) and we have, therefore, again, no reason for imputing the decrease in Small-pox to vaccination. But we may go further than this negative statement, for we have, fortunately, a means of directly testing the alleged efficacy of vaccination. The eleventh Annual Report of the Local Government Board gives a table of the number of successful vaccinations, at the expense of the Poor Rate, in England and Wales, from 1852 to 1881. From the figures of this table I have calculated the numbers in proportion to the population of each year, and have exhibited the result in the dotted line on my Diagram II.; and to this I beg to direct the reader's attention, since it at once dispels some oft-repeated erroneous statements.

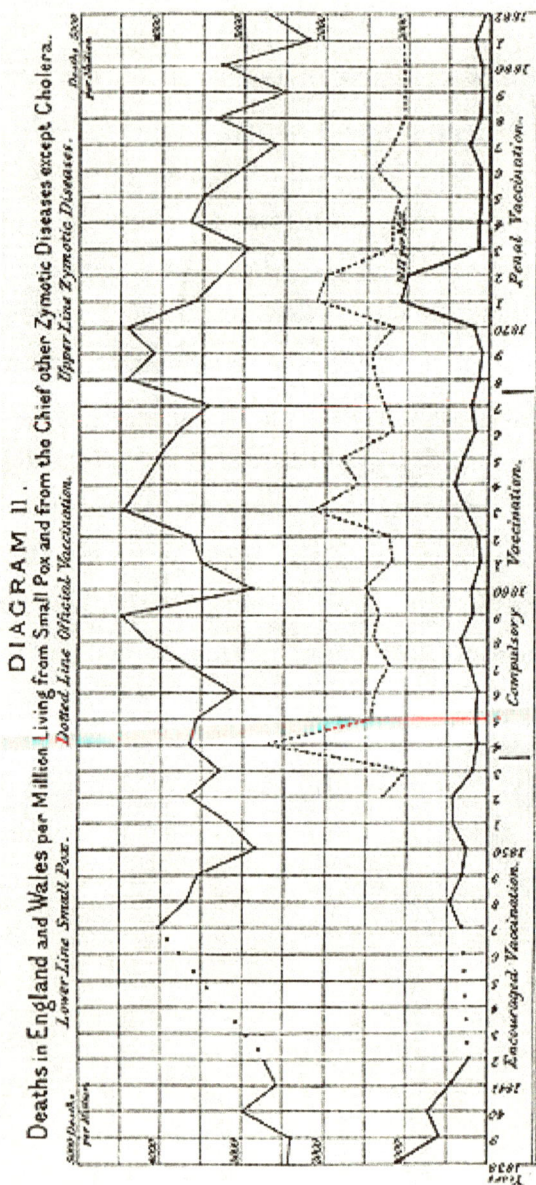

DIAGRAM 11.

Deaths in England and Wales per Million Living from Small Pox and from the Chief other Zymotic Diseases except Cholera.

*Lower Line Small Pox.*    *Dotted Line Official Vaccination.*    *Upper Line Zymotic Diseases.*

12

In the first place we see that, instead of vaccination having increased since the enforcement of penal laws, it has actually diminished; so that the statement so often made by official apologists for vaccination, and repeated by Sir Lyon Playfair in his speech to the House of Commons last year,--that the progressive efficiency of legal vaccination has diminished Small-pox, *is absolutely untrue, since there has been a decrease rather than an increase of "efficient vaccination."*[3] A temporary increase in the number of  vaccinations always takes place during an epidemic of Small-pox, or when an epidemic is feared; but an examination of the curve of vaccination does not support the statement that it checks the epidemic. On careful inspection it will be seen that on three separate occasions a considerable *increase* in vaccinations was followed by an *increase* of Small-pox. Let the reader look at the Diagram, and note that in 1863 there was a very great number of vaccinations, followed in 1864 by an *increase* in Small-pox mortality. Again, the number of vaccinations steadily rose from 1866 to 1869, yet in 1870-71 Small-pox mortality increased; and yet again, in 1876 an *increase* in vaccinations was followed by an *increase* of Small-pox deaths. In fact, if the dotted line showed *inoculation* instead of *vaccination*, it might be used to prove that inoculation caused an increase of Small-pox. I only maintain, however, that it does *not* prove that vaccination diminishes the mortality from the disease. During the panic caused by the great epidemic of 1871-2, vaccinations rose enormously, and declined as rapidly the moment the epidemic passed away, but there is nothing whatever to show that the increased vaccinations had any effect on the disease, which ran its course and then died out like other epidemics.

It has now been proved from the only complete series of official records that exist:--

(1.)--That Small-pox has not decreased so much or so steadily as Typhus and allied fevers.

(2.)--That the diminution of Small-pox mortality coincides with a *diminished*, instead of an increased efficiency of official vaccination.

(3.)--That one of the most severe epidemics of Small-pox on record, within the period of accurate statistics, occurred after 33 years of official, compulsory, and penal vaccination.

These three groups of facts give no support to the assertion that vaccination has diminished Small-pox mortality; and it must always be remembered that we have actually no other extensive body of statistics on which to found our judgment. The utility or otherwise of vaccination is purely a question of statistics. It remains for us to decide, whether we will be guided by the only trustworthy statistics we possess, or continue blindly to accept the dogmas of an interested and certainly not infallible body of professional men, who once upheld inoculation as strongly as they now uphold vaccination.

## Small-pox Has Not Been Mitigated by Vaccination.

It is often asserted that, although vaccination is not a complete protection against Small-pox, yet it diminishes the severity of the disease, and renders it less dangerous to those who take it. This assertion is sufficiently answered by the proof above given, that it has *not* diminished Small-pox mortality; but more direct evidence can be adduced.

The best available records show that, the proportion of deaths to Small-pox cases is the same *now, although a large majority of the population are vaccinated*, as it was a century ago before vaccination was discovered. Dr. Jurin, in 1723; the London Small-pox Hospital Reports, 1746-63; Dr. Lambert, 1763; and Rees' Cyclopædia, 1779; give numbers varying from 16.5 to 25.3 as the per-centage of mortality among Small-pox patients in hospitals;--*the average of the whole being 18.8 per cent.*

Now for the epoch of vaccination. Mr. Marson, 1836-51, and the Reports of the London, Homerton, Deptford,

Fulham, and Dublin Small-pox Hospitals, between 1870 and 1880, give numbers varying from 14.26 to 21.7 as the deaths per cent. of Small-pox patients, *the average being* 18.5. And this, be it remembered, under the improved treatment and hygiene of the nineteenth as compared with the eighteenth century.

These figures not only demonstrate the falsehood of the oft-repeated assertion that vaccination mitigates Small-pox, but they go far to prove the very opposite--that the disease has been rendered more intractable by it; or how can we account for the mortality among Small-pox patients being almost exactly the same now as a century ago, notwithstanding the great advance of medical science and the improvements in hospitals and hospital treatment?[4]

## Small-pox in the Army and Navy.

Here we have a crucial test of the efficacy or uselessness of vaccination. Our Soldiers and Sailors are vaccinated and re-vaccinated in accordance with the most stringent official regulations. They are exceptionally strong and healthy men, in the prime of life, and if vaccination is of any use, Small-pox should be almost unknown among them, and *no soldier or sailor should ever die of it.* They are in fact often spoken of as a "perfectly protected population." Now let us see what are the facts.

A Return has just been issued to the House of Commons, "Small-pox (Army and Navy)," dated "August, 1884," giving the mean strength, the number of deaths from Small-pox, and the ratio per thousand in each service for the twenty-three years 1860-82. An examination of this Return shows us that there has not been a single year without two or more deaths in the Army, and only two years without deaths in the Navy. Comparing the Return on "Vaccination, Mortality," No. 433, issued by the House of Commons in 1877, we find that, in the twenty-three years 1850-72, (the latest there given,) there were many years in which no adult Small-pox deaths were recorded for a number of large towns of from 100,000 to 270,000 inhabitants. Liverpool had none in 3 of the years, Birmingham and Sunderland in 7, Bradford and Sheffield in 8, Halifax in 9, Dudley in 10, while Blackburn and Wolverhampton were each totally without adult Small-pox mortality for 11 out of the 23 years!

It is true that the cases are not strictly comparable, because for these towns we have only deaths of persons aged 20 and upwards given separately, whereas the ages of the Army and Navy range chiefly from about 17 to 45. But, considering the extremely unsanitary state of many of these towns, and their great preponderance in freedom from Small-pox, there is clearly no room left for the alleged effect of *re-vaccination* in securing to our soldiers and sailors immunity from the disease.

But let us now look at the averages for the whole series of years, as affording the best and only reliable test. On working these out carefully, I find the mean Small-pox mortality for the 23 years to be, in the Army 82.96, which we may call 83 per million, and in the Navy 157 per million. Unfortunately no materials exist for an *exact*comparison of these rates with those of the civil population; but with much labour I have made the best comparison I can arrive at. From the Census General Report, 1881, and the Reports of the Registrar-General for the same 23 years as are included in the Army and Navy Return, I have been able to ascertain the Small-pox mortality of males in England and Wales between the years 15 and 55, taken as best representing those of the two services; and the result is a mean Small-pox death rate of 176 per million.[5]

It will be observed that this is but little more than the Navy mortality, though more than double that of the Army, and the question arises, to what is the difference due. And first, why is the Small-pox mortality in the Navy nearly double that of the Army? The regulations as to re-vaccination are the same in both, and are in both rigidly enforced, and the men are pretty equal in stamina and general health. The cause must therefore be in the different conditions of life of the two services; and it seems to me a probable supposition, that the difference arises chiefly from the less efficient ventilation and isolation which are possible on board ship as compared with Army Hospitals.[6]

The general mortality of the Navy from disease appears (from the Registrar-General's Report, 1882, Tables 59 and 65,) to be considerably less than that of the Army, so that the greater mortality from Small-pox must be due to some special conditions. But whatever these are, the conditions of the civil population are certainly much worse. Two-thirds of the families inhabiting Glasgow live in houses of one or two rooms only, and many other towns, including London, are probably not much better. Under such conditions, and with the low vitality induced by insufficient food, overwork, and bad air, we should expect the Small-pox mortality of our civil population to be very much greater than that of the picked class of sailors who enjoy ample food, fresh air, and medical attendance. Where then is the alleged "full security" afforded by re-vaccination, and how are we to characterise the statements circulated at the expense of the public, that "Small-pox is almost unknown in the Army and Navy."[7] If we are to draw a legitimate conclusion from the facts, it is, that the re-vaccination to which our soldiers and sailors are subjected, renders Small-pox more fatal when it attacks them, for thus only can we explain the large mortality among picked healthy men under constant medical supervision, and living under far better sanitary conditions than the mass of the civil population.

One other mode of comparison can be made, showing that even the Army Small-pox death-rate is but little better than that of some large towns, during the same period. The rate per million for the adult population, between the ages 15 and 55, on an average of the years 1860-82 for five very large towns was as follows:--[8]

Manchester, (population 340,211 in 1882), 131 per million.

| Leeds .... | " | 315,998 | " | 119 | " |
|---|---|---|---|---|---|
| Brighton .... | " | 109,595 | " | 114 | " |
| Bradford .... | " | 200,158 | " | 104 | " |
| Oldham .... | " | 115,572 | " | 89 | " |

Of course there are many other towns which have a much higher mortality, but very few are much worse than the Navy. The very worst large town which I can find in the Reports is Newcastle-on-Tyne, which for the same period had an adult Small-pox mortality of 349 per million. But the fact that five of our most populous towns have considerably less adult Small-pox mortality than the Navy, and one of them but little more than the Army, amounts to a demonstration of the uselessness of the most complete re-vaccination.

The general mortality of our adult population is much greater than that of the Army and Navy. From the official sources of information already quoted, I find that the average mortality of the adult population of England, of the ages 15-55, for the years 1860-82, was about 20,000 per million. *[[Editor's Note: Wallace here neglects to take into*

*account that he is only dealing with the adult male population*
*of England; this mistake was corrected in the 1889 edition to*
*reduce the per million rate to 11,300.]]*

That of the Navy, for the same period, was 11,000 per
million from all causes, and only 7,150 from disease.

That of the Army, at home, was 10,300 per million.
Abroad it was nearly double (19,400), but this included all
the deaths from casualties, exposure, &c., in the Abyssinian,
Afghan, Zulu, Transvaal, and other petty wars.

Thus the superior physique of our soldiers and sailors,
together with the sanitary conditions under which they live,
are fully manifested in a mortality from disease only about
half that of the adult civil population of comparable ages.
If we make the same allowance for the influence of these
causes in the case of Small-pox, there remains absolutely
nothing for the alleged protective influence of re-vaccination.

Surely we shall now hear no more of the re-vaccinated
nurses in Small-pox hospitals, (as to whom we have no
statistics, but only vague and usually inaccurate assertions,)
when we have a great, officially recorded experiment to
refer to, extending over 23 years and applied to more than
200,000 men, the results of which directly contradict every
professional and official statement as to the safeguard of re-
vaccination.[9]

# Vaccination Itself A Cause of Disease and Death.

As has been now shown, vaccination is quite powerless either to prevent or to mitigate Small-pox. But this is not all, for there are good grounds for believing that it is itself the cause of much disease and serious mortality.

It was long denied by medical men that syphilis can be communicated by vaccination; but this is now universally admitted, and no less than 478 cases of vaccine-syphilis have already been recorded.[10] But there is also good reason to believe that many other blood-diseases are transmitted and increased by the same means, since there has been for many years a steady increase of mortality from such diseases which is terrible to contemplate. The following table gives the increase of five of these diseases from the Registrar-General's Annual Report for 1880, (page lxxix., Table 34,) and it is very noteworthy that, in the long list of maladies there tabulated, no others, (except Bronchitis, which often follows vaccination though not, probably, transmitted by it,) show any such striking and continuous increase, while the great majority are either stationary or decreasing.

## Annual Deaths in England per Million Living.[11]

| AVERAGE OF 5 YEARS. | 1850-4 | 1855-9 | 1860-4 | 1865-9 | 1870-4 | 1875-9 | 1878-80 |
|---|---|---|---|---|---|---|---|
| Small-pox | 279 | 199 | 191 | 148 | 433 | 82 | 40 |
| Syphilis | 37 | 51 | 64 | 82 | 81 | 86 | 84 |
| Cancer | 302 | 327 | 369 | 404 | 442 | 493 | 510 |
| Tabes Mesenterica | 265 | 261 | 272 | 316 | 299 | 330 | 341 |
| Pyæmia, &c. | 20 | 18 | 24 | 23 | 29 | 39 | 40 |
| Skin Disease | 12 | 15 | 16 | 17 | 18 | 23 | 22 |
| Totals... | 636 | 672 | 745 | 842 | 869 | 971 | 997 |
| Progressive Increase | 0 | 36 | 109 | 206 | 233 | 335 | 361 |

We here see a constant increase in the mortality from each of these diseases, an increase which in the sum of them is steady and continuous. It is true, we have not, and cannot have, direct proof that vaccination is the sole cause of this increase, but we have good reason to believe that it is the chief cause. In the first place it is a *vera causa*, since it directly inoculates infants and adults, on an enormous scale, with whatever blood-disease may exist unsuspected in the system of the infants from whom the vaccine pus is taken. In the next place, no other adequate cause has been adduced for the remarkably continuous increase of these special diseases, which the spread of sanitation, of cleanliness, and of advanced medical knowledge, should have rendered both less frequent and less fatal.

The *increased deaths from these five causes, from 1855 to*

1880, *exceed the total deaths from Small-pox during the same period!* So that even if the latter disease had been totally abolished by vaccination, the general mortality would have been increased, and there is much reason to believe that the increase may have been caused by vaccination itself.

# PART II. COMPARATIVE MORTALITY OF THE VACCINATED AND THE UNVACCINATED.

In his speech in the House of Commons, June 19th, 1883, Sir Lyon Playfair made the following statement:--"An analysis of 10,000 cases in the Metropolitan Hospitals shows that 45 per cent. of the Unvaccinated patients die, and only 15 per cent. of Vaccinated patients;" and he further showed that statistics of a similar character had been published in other countries. It will no doubt be objected by my readers that these statistics, if correct, are a complete proof of the value of vaccination; and I shall be expected to show that they are incorrect or give up the whole case. This I am prepared to do; and I now undertake to prove--firstly, that the figures here given are unreliable; and, secondly, that such statistics *necessarily* give false results unless they are classified according to the age-periods of the patients.

## The Per-centages of Vaccinated and Unvaccinated Unreliable.

The simple fact of death from Small-pox is easily ascertained, and has been for many years accurately recorded.

But, whether the deceased person had been vaccinated or not, is a fact by no means easily ascertained, because confluent Small-pox (which alone is ordinarily fatal) obliterates the vaccination marks in most cases, and the death is then usually recorded among the unvaccinated or the doubtful. For this reason alone the official record--*vaccinated* or *unvaccinated*--is altogether untrustworthy, and cannot be made the subject of accurate statistical enquiry.

But there are other reasons why the comparison of the deaths of these two classes is worthless. Deaths registered as *unvaccinated* include--

(1.)--Infants dying under vaccination age, and who, therefore, have no corresponding class among the vaccinated, but among whom the Small-pox mortality is greatest.

(2.)--Children too weakly or diseased to be vaccinated, and whose low vitality renders any severe disease fatal.

(3.)--A large but unknown number of the criminal and nomad population who escape the vaccination officers. These are often badly fed and live under the most unsanitary conditions; they are, therefore, especially liable to suffer in epidemics of Small-pox or other zymotic diseases.

It is by the indiscriminate union of these three classes, together with those erroneously classed as unvaccinated owing to the obliteration of marks or other defect of evidence, that the number of deaths registered "*unvaccinated*" is swollen far beyond its true proportions, and the comparison with those registered "*vaccinated*" rendered altogether untrustworthy and misleading.

This is not a mere inference, for there is much direct evidence that the records "unvaccinated" and "no statement" in the Reports of the Registrar-General are often erroneous. As the chief argument for vaccination rests upon this class of facts, a few examples of the evidence referred to must be here given.

(1.)--Mr. A. Feltrup, of Ipswich, gives a case of a boy aged 9, who died of Small-pox, and was recorded in the certificate as "unvaccinated." By a search in the register of successful vaccinations it was found that the boy, Thomas Taylor, had been successfully vaccinated on the 20th May, 1868, by W. Adams. (*Suffolk Chronicle*, May 5, 1877.)

(2.)--In "Notes on the Small-pox Epidemic at Birkenhead, 1877." By Fras. Vacher, M.D., (.,) we find the following:--

"As regards the patients admitted to the fever hospital or treated at home, those entered as vaccinated displayed undoubted cicatrices, as attested by competent medical witnesses, and those entered as not vaccinated were admitted unvaccinated or without the faintest mark. *The mere assertions of patients or their friends that they were vaccinated counted for nothing, as about 80 per cent. of the patients entered in the third column of the table ('unknown') were reported as having been vaccinated in infancy.*" (The italics are my own.)

(3.)--Bearing upon this important admission, we have the following statement in Dr. Russell's Glasgow Report, 1871-2 ():--

"Sometimes persons were said to be vaccinated, but no marks could be seen, very frequently because of the abundance of the eruption. In some cases of those which recovered, an inspection before dismissal discovered vaccine marks, sometimes 'very good.'"

(4.)--In the Registrar-General's Return of Births and Deaths for London, for the week ending Oct. 13th, 1883, three Small-pox deaths are recorded from the Metropolitan Asylums Hospital at Homerton, and they are stated to be of three unvaccinated children--one, four, and nine years respectively--all from 3, Medland Street, Stepney. Thereupon the mother of two of the children made a declaration that the return was *not true*. She states:--"All my five children were beautifully vaccinated. My three sons were attacked with Small-pox; the two youngest are dead, the eldest is better." (Signed) Ann Elizabeth Snook, 3, Medland Street, Stepney, October 23rd, 1883. A full account of this case was published in the *Bedfordshire Express* by Mr. A. Stapley, and also in the *Vaccination Enquirer* of December, 1883.

(5.)--In 1872, Mr. John Pickering, of Leeds, carefully investigated a number of cases entered as "not vaccinated" by the medical officers of the Leeds Small-pox Hospital, tracing out the parents, examining the patients if alive, or obtaining the certificate of vaccination if they were dead. The result was, that 6 patients, entered as "not vaccinated," and still living, were found to have good vaccination marks; while 9 others who had died, and whose deaths had been registered as "not vaccinated," were proved to have been successfully vaccinated. In addition to these, 8 cases were proved to have been vaccinated, some of them three or four times, but unsuccessfully, and 4 others were certified "unfit to be vaccinated," yet all were alike entered as "unvaccinated." The full particulars of this investigation are to be found in a pamphlet by Mr. Pickering, published by F. Pitman, 20, Paternoster Row, London.

(6.)--As further corroborative evidence of the untrustworthiness of all records on the subject emanating from medical men, the following quotation from an article on "Certificates of Death," in the *Birmingham Medical Review* for January, 1874, is important; the italics are my own:--"In certificates given by us voluntarily, and to which the public have access, it is scarcely to be expected that a medical man will give opinions which may tell against or reflect upon himself in any way. In such cases he will most likely tell the truth, *but not the whole truth*, and assign some prominent symptom of the disease as the cause of death. As instances of cases which may tell against the medical man himself, I will mention *erysipelas from vaccination*, and puerperal fever. A death from the first cause  occurred not long ago in my practice, and although I had not vaccinated the child, yet *in my desire to preserve vaccination from reproach, I omitted all mention of it from my certificate of death.*"

The illustrative facts now given cannot be supposed to be exceptional, especially when we consider the great amount of time and labour required to bring them to light; and taken in connection with the astounding admissions of medical men, of which examples have been just given, they prove that *no dependence can be placed on the official records of the proportions of vaccinated and unvaccinated among Small-pox patients*; while, if Mr. Vacher's method of registration is usually followed, about 80 per cent. of those classed by the Registrar-General under the heading "no statement" have been really stated, by their parents or friends, to have been vaccinated.

## Our Hospital Statistics Necessarily Give False Results.

But a still more serious matter remains to be considered, and it is a striking proof of the crude and imperfect evidence on which the  important question of the value of vaccination has been decided, that the point in question has been entirely overlooked by every English advocate of vaccination, although it involves an elementary principle of statistical science.

This point is, that even if the records in our hospitals, "vaccinated" and "unvaccinated," were strictly correct, it can be demonstrated that true results cannot be deduced from them without a comparison of *the mortality of the vaccinated and the unvaccinated at corresponding ages*, and this information our official returns do not give.

The requisite comparison has, however, been made on a population of about 60,000, consisting of the officials and workmen employed on the Imperial Austrian State Railways, by the Head Physician, Dr. Leander Joseph Keller; and his results during the years 1872-3 are so important that it is necessary to give a brief abstract of them.[12]

(1.)--It is shown that the death-rate of Small-pox patients is greatest in the first year of life,  then diminishes gradually to between the 15th and 20th year, and then rises

31

again to old age; *thus following exactly the same law as the general mortality.*

(2.)--The Small-pox death-rate, among over 2,000 cases, was 17.85 per cent. of the cases, closely agreeing with the general average. That of the unvaccinated was 23.20 per cent., while that of the vaccinated was only 15.61 per cent.

(3.)--This result, apparently so favourable to vaccination, is shown to be wholly due to the excess of the unvaccinated in the first two years of life,[13] and to be *a purely numerical fact entirely unconnected with vaccination.* This is proved as follows:--Taking, first, all the ages above 2 years, the death-rates of the vaccinated is 13.76, and of the unvaccinated 13.15,--almost exactly the same, but with a slight advantage to the unvaccinated.

Taking now the first two years, the death-rate is found to be as follows:--

|  | Vaccinated. | Unvaccinated. |
| --- | --- | --- |
| First year of life . . . | 60.46 | 45.24 |
| Second year of life . . . | 54.05 | 38.10 |

Thus the Small-pox death-rate is actually *less* for the unvaccinated than for the vaccinated in infants, and *equal* for all the higher ages; yet the average of the whole is higher for the unvaccinated, *simply on account of the greater proportion of the unvaccinated at those ages at which the mortality is universally greatest.*

It is thus made clear that any comparison of the Small-pox mortality of the vaccinated and the unvaccinated, *except at strictly corresponding ages*, leads to entirely false conclusions.

This curious and important fact may perhaps be rendered more easily intelligible by an illustration. Let us take the whole population up to 20 years of age, and divide it into two groups--those who go to school, and those who do not. If the Small-pox mortality of these were separately registered, it would be found to be very much greater among the non-school goers,--composed chiefly of infants, and of children too weakly to be sent to school, amongst whom the mortality is always very great, so much so that a doctor of wide experience--Dr. Vernon, of Southport--has stated that, he had never known an infant under one year of age recover from Small-pox. But we should surely think a person either silly or mad who argued from such statistics that school-going was a protection against the disease, and that school children formed a "protected population." Yet this is exactly comparable with the reasoning of those who adduce the greater mortality among unvaccinated Small-pox patients of all ages, as the very strongest argument in favour of vaccination![14]

Good statistics and good arguments cannot be upset, or even weakened, by those which are bad. I have now shown that the main argument relied on by our adversaries, rests on thoroughly unsound statistics, inaccurate to begin with, and wrongly interpreted afterwards. Those which I have used, on the other hand, if not absolutely perfect, are yet the best and most trustworthy that exist. I ask statisticians and men of unbiassed judgment to decide between them.

## Conclusion from the Evidence.

The result of this brief enquiry may be thus summarized:--

(1.)--Vaccination does not diminish Small-pox mortality, as shown by the 45 years of the Registrar-General's statistics, and by the deaths from Small-pox of our "re-vaccinated" soldiers and sailors being as numerous as those of the male population of the same ages of several of our large towns, although the former are picked, healthy men, while the latter include many thousands living under the most unsanitary conditions.

(2.)--While thus utterly powerless for good, vaccination is a certain cause of disease and death in many cases, and is the probable cause of about 10,000 deaths annually by five inoculable diseases of the most terrible and disgusting character, which have increased to this extent, steadily, year by year, since vaccination has been enforced by penal laws!

(3.)--The hospital statistics, showing a greater mortality of the unvaccinated than of the vaccinated, have been proved to be untrustworthy; while the conclusions drawn from them are shown to be necessarily false.

If these facts are true, or anything near the truth, the enforcement of vaccination by fine and imprisonment of unwilling parents, is a cruel and criminal despotism, which it behoves all true friends of humanity to denounce and oppose at every opportunity.

Such legislation, involving as it does, our health, our liberty, and our very lives, is too serious a matter to be allowed to depend on the misstatements of interested officials or the dogmas of a professional clique. Some of the misstatements and some of the ignorance on which you have relied, have been here exposed. The statistical evidence on which alone a true judgment can be founded, is as open to you as to any doctor in the land. We, therefore, demand that you, our representatives, shall fulfil your solemn duty to us in this matter, by devoting to it some personal investigation and painstaking research; and if you find that the main facts as here stated are substantially correct, we call upon you to undo without delay the evil you have done.

WE, THEREFORE, SOLEMNLY URGE UPON YOU THE IMMEDIATE REPEAL OF THE INIQUITOUS PENAL LAWS BY WHICH YOU HAVE FORCED UPON US A DANGEROUS AND USELESS OPERATION--AN OPERATION WHICH HAS ADMITTEDLY CAUSED MANY DEATHS, WHICH IS PROBABLY THE CAUSE OF GREATER MORTALITY THAN SMALL-POX ITSELF, BUT WHICH CANNOT BE PROVED TO HAVE EVER SAVED A SINGLE HUMAN LIFE.

*Notes Appearing in the Original Work*

[1]From the Registrar-General's Annual Summary of Deaths, &c., in London, 1882. Table 23, p. xxv.

[2]From the Registrar-General's Annual Report, 1882. Table 32, p. lxiii.

[3]It is curious that even the Registrar-General appears to be ignorant of the fact that, official vaccination has not increased in efficiency since the penal laws came into force. In his Report for 1880, p. xxii., he says--"These figures show conclusively that, *coincidently with the gradual extension of the practice of vaccination*, there has been a gradual and notable decline in the mortality from Small-pox at all ages." As, however, there has not been shown to have been any such "gradual extension of the practice of vaccination," but, so far as official records go, just the reverse, the whole argument falls to the ground! It is true that this curve does not exhibit

the numbers of the vaccinated population, which there is no means of arriving at. Dr. Seaton, in his evidence before the Parliamentary Committee in 1871, stated that, before 1853 the average vaccinations were 31.8 per cent. of the births, and in the ten years 1861-70, 49.46 per cent. These are public vaccinations, but they probably include the bulk of the whole; and the figures seem to show that the proportion of the population vaccinated is much less than is usually supposed.

[4]The following authorities have been examined for the facts and figures of this section.

Dr. Jurin (18,066 cases) and Dr. Lambert (72 cases) given in "Analyse et Tableau de l'Influence de la Petite Verole; par E. E. Duvillard. Paris, 1806." (p, 113.) London Small-pox Hospitals (6,454 cases) given in "An account of the Rise, Progress, and State of the Hospitals for relieving poor people afflicted with the Small Pox, and for Inoculation," appended to "A Sermon preached before the President and Officers of the Hospital . . . . by the Bishop of Lincoln. London, 1763."

Rees' Cyclopædia, 1779, Vol 2, Art. Inoculation Col. INP. par. 5, (extract). "From a general calculation it appears that, in the Hospitals for Small Pox and Inoculation, 72 die out of 400 patients having the distemper in the natural way." Total cases before Vaccination, 24,994. Mr. Marson, Resident Surgeon to the Small-pox and

Vaccination Hospital, London, (5,652 cases); given in the Blue Book on The History and Practice of Vaccination, 1857, . London Hospitals, 1870-72, (14,808 cases); in the Report of a Committee of the Managers of the Metropolitan Asylum District, July 1872, . London Hospitals, 1876-80, (15,172 cases); in a letter to *The Times* of November 8th, 1879, from W. F. Jebb, Clerk to the Metropolitan Asylum District. Homerton, (5,479 cases); from the Report of the Committee, 1877.

Deptford, (3,185 cases); from the Report of the Medical Superintendent, 1881.

Fulham, (1,752 cases); from the Report of the Medical Superintendent, 1881.

Dublin, (2,404 cases); from the Annual Report of the Committee, 1880.

Total cases after Vaccination, 48,451.

The extracted figures and per-centages have been all carefully verified, and the averages have been obtained by dividing the total number of deaths multiplied by 100, by the total number of cases.

[5]The following are the data on which this calculation is founded:--

In the General Report of the last Census, Table 14, , the numbers of males at successive ages are given for the three last Censuses--1861, 1871, and 1881. By a simple calculation it is found that the number of males of all ages is to that of males aged 15-55 in the proportion of 1 to .528. Table 4., of the same Census Report, gives the male population for the middle of each of the 23 years included in the Army and Navy Return. The mean of these numbers is 11,167,500; and this sum, multiplied by the factor .528, gives 5,896,500 for the average male population of the ages 15-55 for those years. From the tables of "Causes of Death at different Periods of Life" in the twenty-three successive Reports of the Registrar-General, 1860-1882, I have extracted the deaths from Small-pox of males aged 15-55, the mean annual value of which is 1,041; and this number, divided by the number of millions in the corresponding population (5.8965), gives the death-rate per million = 176. The limit of age, 15-55, has been taken because the General Report of the Census of 1881, Table 40, gives, for the Army and Navy, 7,530 men over 45, and 28,834 under 20 years of age.

[6]An Officer of the Royal Marine Artillery, of great experience, confirms this view. He assures me that isolation is absolutely impossible on board a ship of war. But if this is the explanation of the phenomenon, it is itself a proof of the complete inefficacy of re-vaccination, which not only does not protect men from *catching* Small-pox, but allows them to *die* of it quite as much as--and, allowing something

39

for the superiority of sanitation, even more than--the adult civil population, only partially vaccinated and hardly ever re-vaccinated!

[7]The following are a few of these assertions. The italics are to call attention to the essential words of each statement. The "Lancet," of March 1st, 1879, says:-- "Vaccination needs to be repeated well once in a lifetime, *and then the immunity is almost absolute.*" The Medical Officer of the General Post Office says, in a circular dated June, 1884--"The *only means* of securing protection against Small-pox is by re-vaccination . . . . it is desirable, *in order to obtain full security*, that the operation should be repeated at a later period of life." In the tract on "Small-pox and Vaccination" issued by the National Health Society, and now being widely circulated at the expense of the ratepayers, with the sanction of the Local Government Board, we find this statement:--"Every Soldier and Sailor is re-vaccinated; the result is that *Small-pox is almost unknown in the Army and Navy,* even amid surrounding epidemics." The above statements are proved by the Official Returns now issued to be absolutely untrue, and must have been ignorantly and recklessly made without any adequate basis of fact.

[8]These figures have been thus obtained:--the Registrar-General's Summary, 1882, (Table 7, p. xv.) gives the Small-pox

deaths per 1,000, for twenty great Towns, for the years 1872-82. The Parliamentary Return, "Vaccination, Mortality," 1877, gives the Small-pox mortality and population of a considerable number of towns for the years 1847-72. From these two official papers the Small-pox mortality per million of the whole male population from 1860 to 1882, for such towns as occur in both the tables, is easily obtained. The average Small-pox death-rate for all England is found to be 211.7, while that of the ages 15-55 is 176. These numbers are in the proportion of 1 to .83; hence the total Small-pox mortality of any town multiplied by the factor .83 will give, approximately, the mortality at ages 15-45. The proportion has been obtained from males only, but that of the two sexes combined will not be materially different.

[9]It will, perhaps, be objected that the Army and Navy are in great part stationed abroad and even in the tropics. But this renders our case all the stronger, because Small-pox is *less prevalent* in the tropics. In the Blue Book on Vaccination, prepared by the General Board of Health in 1857, it is stated, as the mean result of many years records, that, "the deaths have been four times as numerous among troops in the United Kingdom as in temperate colonies, and eight times as numerous as in tropical colonies," (.) It follows, that had the home force alone been given in the Return, the Small-pox mortality of the revaccinated and "completely protected" troops would probably have been

*greater* than that of the civil population of corresponding ages!

[10]See Mr. Tebb's "Compulsory Vaccination in England," , (Note,) for a list of the authorities for these cases.

[11]This Table has not been continued in later Reports; but we find in 1882 that Cancer (the only disease of the five separately tabulated) goes on steadily increasing, the mortality given being, for 1880, 514; for 1881, 522; and for 1882, 532 per million!

[12]Report on Small-pox cases among the Employés of the Imperial Austrian State-Railway Company for the year 1873. Translated from the German by Mrs. Hume-Rotheray. National Anti-Compulsory Vaccination League.

[13]This applies to Austria. In England vaccination is usually performed earlier, yet, in a pamphlet entitled "*Plain Facts on Vaccination*," by G. Oliver, about 1872, it was stated that in the Small-pox Hospital, Hampstead,--"The number of the unvaccinated patients, up to the age of ten years, greatly preponderates over the vaccinated of corresponding ages." In the Homerton Small-pox Hospital in the eight years 1871-77, there were 147 unvaccinated patients under 2 years old, to 20 vaccinated, including among these the doubtful cases.

[14]It seems incredible, but is nevertheless a fact, that in the whole body of official statistics and reports relating to Vaccination, from the elaborate "History and Practice in

Vaccination" in 1857, containing Mr. Marson's celebrated paper on the Statistics of the London Small-pox Hospitals, down to the latest Hospital Statistics quoted by Sir Lyon Playfair, there is no recognition whatever of the necessity of comparing *corresponding ages* in order to obtain *true results* as to the *comparative mortality of the vaccinated and unvaccinated*. Yet it was largely owing to these very Hospital Statistics that the Penal Vaccination Laws were passed, and have been since upheld by Parliament!

\* \* \* \* \*

# FORTY-FIVE YEARS' REGISTRATION STATISTICS.
## A CORRECTION. (S509: 1895)

Sir,--While thanking my friend Mr. Alex. Wheeler for his too complimentary references to the little I have done for the cause of freedom as regards the tyranny of the Vaccination laws, I wish to make a remark as to one portion of his article which conveys an erroneous impression. Mr. Wheeler says that he could not agree with my conclusion that "Vaccination may have caused more deaths than smallpox itself." This I am not surprised at, because I do not myself accept such a statement, which is certainly not mine. My words, carefully chosen, are--"an operation which has admittedly caused many deaths, which is probably the cause of greater mortality than smallpox itself"--and I call attention to the change from the past tense in the first part of the passage to the present tense--"is probably the cause"--in the latter part. This clearly means, not that "Vaccination may have caused more deaths than smallpox"--as Mr. Wheeler states it, without any limitation of time, which would of course be an absurdity--but that, at the *present time*, as the result of general Vaccination for about fifty years, it may *now* be the cause of more deaths than smallpox. This conclusion is drawn from the table of the steadily-increasing mortality from certain inoculable diseases (page 24 of my pamphlet), which *increase*, in thirty years (1850-1880), was 357 per million (an increase which has continued since), while the deaths from smallpox have not, for many years, averaged more than *one-fifth* of this amount. If, therefore, only *one-fourth part* of the large and steady increase of these diseases is due to Vaccination, then my belief that Vaccination *is now* the cause of greater mortality than smallpox itself is fully justified; and in the contention that this is "probably" the case I do not think that I shall find myself in the minority among the readers of the *Inquirer*. This indirect effect of Vaccination is further increased by its direct effects, which are now known to be far more terrible, and to produce far greater mortality than was formerly suspected or admitted.

I wish to take the opportunity of requesting such of your readers as may have copies of my pamphlet to erase from line 11 on page 21, to line 9 on page 22, 2nd edition (or, in the first edition, from line 8 on page 20 to line 4 on page 21--Ed. V. I.), as the figures and conclusions therein are erroneous.

www.ingramcontent.com/pod-product-compliance
Lightning Source LLC
Chambersburg PA
CBHW030838300326
41935CB00037B/642